## Contents

**About the Author**

**Preface**

**1. Introduction and Discovery of the Microbial World**

**2. Microbial Taxonomy**

**3. Eukaryotic Cell Organelles**

**4. Journey to the Microbial World**

**5. Bacteria**

**6. Archaea**

**7. Microbial Genetics of Bacteria**

**8. Overview of Viruses**

**9. Mycoplasma**

**10. Genetic Engineering**

# Chapter 1
# INTRODUCTION AND DISCOVERY OF THE MICROBIAL WORLD

Welcome to microbiology – the study of microorganisms. Microorganisms are so small that it is invisible to naked eye. They are distributed all over the world and even inside the human body. Microorganisms or microbes included a massive range of organisms including bacteria, fungi, viruses, algae, archaea and protozoa. So, microorganisms are microscopic but not cellular. Microbes contain 50% of the biological carbon and 90% of the biological nitrogen on Earth. Most importantly, certain microorganisms carry out photosynthesis. What is microbiology? Microbiology is the foundation of biological source. This included eukaryotes such as fungi and protists and prokaryotes such as bacteria and certain algae. Viruses are also included. A scientist who specializes in the area of microbiology is called a microbiologist. Microbiologist pursue careers in many fields, including agricultural, environmental, food and industrial microbiology; public health, resource management; basic research, education and pharmaceuticals.

Although microorganisms have existed for a long time, their existence was unknown until the invention of microscope in the 17th century. Jensen (1950) was the first person who magnified objects with hand lens. Later Antoni van Leeuwenhoek (1863) of Holland an amateur lens grinder, was the person to make glass lens and observed microorganisms. Antoni van Leeuwenhoek is the first scientist to perform work in the field of microscopy and for his contributions towards the establishment of microbiology. The year 1674, marks the birth of microbiology when Antoni van Leeuwenhoek looked at a drop of lake water through a glass lens and found microscope creatures and named them as animalcules. Antoni van Leeuwenhoek (24 october 1632 - 26 August 1732) was a dutch businessman and scientist. Leeuwenhoek, a son of craftsman, was born in Delft in 1632 and died in 1732, at the age of 91 years. At the age of 22 , he married to Barbara. Using his primitive microscope he examined many substances, including pond water, soil, blood and saliva, and found many microscopic creatures within them. He thought that these creatures were small animals and therefore called them animalcules or little animals. They later come to be known as microorganisms or microbes. For this important contribution made in the identification of microorganism, Antoni van Leeuwenhoek is known as the **Father of Microbiology**

1.1 Discovery of microorganism

Earth is 4.6 billion years old. Scientist have evident that cells first appeared on Earth between 3.8 billion years ago; these organism were exclusively microbial. In fact microorganisms were the only life on Earth for most of its history. The discovery of the microbial world immediately raised question regarding the origin of microorganisms. Living organisms such as plants and animals do not originate spontaneously. However some belive that these microorganisms arose spontaneously and this theory came to be known as the theory of spontaneous generation or abiogenesis. The basic idea of spontaneous generation can easily be understood. For example, if food is allowed to stand for some time, it putrefies. When the putrefied material examined microscopically, it is found to be teeming with bacteria. Some people said they arose spontaneously from nonliving materials that is spontaneous generation. So, people had belived in generation- that living organism could develop from nonliving matter. But in 1748, the English priest John Needham (1713-1781) reported the results of his experiment on spontaneous generation. Needham boiled mutton broth infused with plant or animal matter, hoping to kill all pre-existing microbes. He then sealed the flask. After a few days, Needham observed that the broth had become cloudy and contained microorganisms. He argued that the new microbes must have arisen spontaneously. Needham (1748) put the question to an experimental test. He wrote: "For my purpose therefore I took a quantity of mutton-gravy hot from the fire, and shut it up in a phial, closed with a cork so well masticated, that my precautions amounted to as much as if I had sealed my phial hermetically. I thus effectually excluded the exterior air, that it may not be said my moving bodies drew their origin from insects, or eggs floating in the atmosphere. I would not instill any water, lest, without giving it as intense a degree of heat, it might be thought these productions were conveyed through that element. My phials swarmed with life."

Lazzaro spallanzani (1729-1799) did not agree with Needham conclusions. However he performed experiments. Lazzaro spallanzani put broth into four flasks. Flask 1 was left open and flask 2 sealed. Flask 3 was boiled and then left open. Flask 4 was boiled and then sealed. Then he observed out of 4 flasks, flasks 1, 2 and 3 were found microbes but flask 4 does not found any microbes. In Lazzaro spallanzani's experiment he proved microorganisms could be killed by boiling.

Today spontaneous generation is generally accepted to have been decisively dispelled during the 19th century by the experiment of Louis Pasteur. Pasteur prepared a nutrient broth similar to the broth one would use in soup. Then, he placed equal amount of the broth into two long nacked flasks. He left one flask with a straight neck. The other he bent to form an S shape. Then he boiled the broth in each flask to kill any living matter in the liquid. The sterile broths were then left to sit, at room temperature and exposed to the air, in their open mouthed flasks. After several weeks, Pasteur observed that the broth in the straight-neck flasks was discoloured and cloudy. Other broth in the curved neck flask had no changed. He concluded that germ in the air were able to fall unobstructed down the straight neck flask and contaminate the broth. Pasteur experiment showed that microbes cannot arise from nonliving materials under the conditions that existed on Earth during his lifetime. But his experiment did not proved that spontaneous generation never occurred. Another contribution of Louis pasteur to germ theory of disease. Germ theory of disease transmission was established by Pasteur. It states that microorganisms known as pathogens or germs can lead to disease. Germs may refer to any type of microorganism that can cause disease such as protists, fungi, viruses, prions or viriods. Pasteur is regarded the **Father of bacteriology and pasteurization**.

## 1.2 Koch Postulates

Microorganism could cause disease. Even in the sixteenth century it was thought that a disease could be transmitted from a diseased person to healthy person, after the discovery of microorganism, it was widely believed that they were responsible, but definite proof was lacking. Heinrich Hermann Robert Koch (11 December 1843-27 May 1910) was a microbiologist. Koch used mice as experimental animals. While studying the anthrax disease in cattle, Koch had found rod shaped organisms in the blood of the diseased animals. He injected healthy mice with the blood from a diseased animal. He found that the disease was transmissible to the mice. Next, he removed a piece of the spleen from the diseased mice and transferred it to sterile serum and obtained good growth of the rod shaped organism. By injecting these organisms into healthy animals he was able to reproduce the disease. From these studies and his sequential observation lead t the establishment of postulates known as koch's postulates. Koch postulates are four criteria designated to establish a causative relationship between a microbe and a disease. Koch's postulates are the following:

1. The microorganism must be found in abundance in all organism suffering from the disease, but should not be found in healthy organisms.

2. The microorganism must be isolated from a diseased organism and grown in pure culture.

3. The cultured microorganism should cause disease when introduced into healthy organism.

4. The microorganism must be reisolated from the inoculated, diseased experimental host and identified as being identical to the original specific causative agent.

**Table 1.1 Three hundred years of microbiology: some discovery in microbiology, 1684-1876**

| Year | Investigator(s) | Discovery |
|---|---|---|
| 1684 | Antoni van Leewenhoek | Bacteria |
| 1798 | Edward Jenner | Smallpox vaccination |
| 1857 | Louis Pasteur | Microbiology of lactic acid fermentation |
| 1860 | Louis Pasteur | Role of yeast in alcoholic fermentation |

| 1864 | Louis Pasteur | Fallacy of spontaneous generation |
| 1867 | Robert Lister | Antiseptic principles of surgery |
| 1876 | Ferdinand Cohn | Endospores |

## 1.3 Application of microorganisms

The term microorganisms include bacteria, fungi, viruses and protozoa. Microbiology is the study of microorganisms, microscopic organisms that include in particular the bacteria, a large group of very small cells that enormous basic and practical significance (Madigan et al. 2015). Microbiology composed of several subdiscipline such as biotechnology, immunology, microbial ecology, microbial genetics, microbial physiology, microbial systematic, molecular biology, virology etc. Microorganisms are widely distributed on the biosphere, because of their metabolic activity and they can easily grow in a wide range of environmental condition. Microorganisms play an important role in our lives. Some of them are beneficial in many ways whereas some others are harmful and cause disease.

**Microorganisms and medicine**

Currently, the use of microbes is not limited to the use in vaccination. Microorganisms are widely used in medicine. A common use of bacteria in modern medicineis as a delivery capsule for toxic drugs. A delivery capsule was a necessary advancement in the treatment of cancer. Lastly, Dr. Ifat Rubin-Bejerano started a company name Immunexcite which seeks to deveop a cancer therapy.

**Microorganisms in bioremediation**

Microorganisms can also be used to help clean up pollution created by human activities, a process called bioremediation. Bioremediation is involved in degrading, removing various chemicals and physical wastes from the environmental through the action of bacteria, fungi and plants. It also helps in complete destruction of the pollutants.

**Microorganisms and agriculture**

Modern agricultural practices largely rely on high inputs of mineral fertilizers to high yields and involve applications of chemical pesticides to protect crop from disease. Some of the commonly promoted and used beneficial microorganisms in agriculture worldwide include Rhizobia, Mycorrhiza, Azospirillum, bacillus, Pseudomonas, Trichoderma, Streptomyces species and many more.

### Exploration of microbes for Omega-3 fatty acid production

Earth at almost 5 miles per second. Microbes in the major component on this earth. Microbes is a living thing.Microbes are found everywhere in our world. The microbial world is the largest reservoir in biodiversity. Microbes usually lives in community. Over 99% of microbes constriute to the quality of human live. Microbes are essential for our lives. If we look we will find the microbes here and there. Omega -3-fatty acids (polyunsaturated fatty acids) are known as essential fatty acid. Eicosepentaenoic acid and docosahexaenoic acids are typical of Omega-3 fatty acids, sources are derived from cold water fish oils. Recently, most common algae or algae like microorganism used for the production of DHA. DHA is being helpful in the development of large brain. Microorganisms have potential to produce commercially useful materials. Microbes is a single celled (unicellular) or in a colony of cells. Antonie Van Leeuwenhoek was first to study microorganisms. Microorganisms are produce to docoshaexanoic (DHA) and eicosapentaepic (OEPA) w-3 fatty acids. Antonie Van Leeuwenhoek (The father of microbiology) was first to study microorganisms. Omega-3 fatty acids are a large group of fats called alpha-linoleic acid. Not only, microalgae but also thraustochytrids to the production of omega-3 PUFAs. Human can get alpha linolenic acids from plants.

Exploration is the act of searching for the purpose of discovery. The application of Omega-3 fatty acids for human health are rapidly expending. Omega-3 fatty acids play an important role in maintaining the earth ecosystem. Microbes consist of bacteria, Archaea, protozoa, algae, fungi, viruses and multicellular animal parasite, (helminths). Research has been performed to maximize microalgal production of Omega-3 fatty acid. This Omega-3-biotechnological process is green and sustainable process for the omega 3 fatty acid production.

## References

1: A. E. (n.d.). The Spontaneous Generation theory was disproved by Spallanzani, not by Pasteur., 1–14. doi: 10.13140/RG.2.2.33427.17441

2: Adams, E. A.; "Lazzaro Spallanzani", The Scientific Monthly, 1929, 29(6):528-537.

3: Clin Microbiol Rev. 1996 Jan;9(1):18-33.

4: Rev Argent Microbiol. 2010 Oct-Dec;42(4):311-4. doi: 10.1590/S0325-75412010000400013.

5: Journal of Microbiology Research 2014, 4(3): 148-151 DOI: 10.5923/j.microbiology.20140403.04

6: Abatenh E, Gizaw B, Tsegaye Z, et al. Application of microorganisms in bioremediation-review Journal of Environmental Microbiology December 2017;1(1):02-09.

7: Ji X-J, Ren L-J and Huang H (2015) Omega-3 biotechnology: a green and sustainable process for omega-3 fatty acids production. Front. Bioeng. Biotechnol. 3:158. doi: 10.3389/fbioe.2015.00158

8: ISCA Journal of Biological Science ISSN 2278-3202 Vol. 1(3), 78-83, July (2012)

9: Swanson, D.; R. Block, and S. A. Mousa. 2012. Omega-3 fatty acids EPA and DHA: health benefits throughout life. Advances in Nutrition (Bethesda, Md.). 3:1-7. doi: 10.3945/an.111.000893.

10. https://doi.org/10.1002/biot.201500279

11. Puri M., Thyagarajan T., Gupta A., Barrow C.J. (2015) Omega-3 Fatty Acids Produced from Microalgae. In: Kim SK. (eds) Springer Handbook of Marine Biotechnology. Springer, Berlin, Heidelberg

12. The Professional Animal Scientist · June 2009 DOI: 10.15232/S1080-7446(15)30713-0

13. doi: 10.1093/femsle/fnv166 Advance Access Publication Date: 14 September 2015

14. Indian Journal of Marine Sciences Vol. 35(4), December 2006, pp. 359-363

# Chapter 2

# Microbial Taxonomy

Taxonomy (Greek, taxis=arrangement, nomous=law or rule) is the science that dealing with the description, identification, naming, and classification of the organism. Carolus Linnaeus is called the Father of Taxonomy.

Classification: Classification is the arrangement of organisms into groups (taxa) on the basis of relationships.

Nomenclature: Nomenclature is the names of taxonomic groups according to the International Code of Nomenclature of Bacteria.

Identification: To identify of taxon.

Taxonomic ranks: The highest rank is called domain. Several levels or ranks are used in bacterial classification.

Table 1. Taxonomic ranks

| Rank | Example |
|---|---|
| Domain | Bacteria |
| Phylum | Proteobacteria |
| Class | gama-proteobacteria |
| Order | Enterobacteriales |
| Family | Enterobacteriaceae |
| Genus | Escherichia |
| Species | E. coli |

**The Evolutionary Tree of Life**

Evolution is change in the heritable characteristics of biological population over generations. The scientific theory of evolution by natural selection was proposed by Charles Darwin and Alfred Russel Wallace in mid 19th century and was set out in detail in Darwin's book on the Origin of Species (1859).Evolution occurs in any self replicating system in which variation occurs as the result of mutation.

**The three Domains of life**

The evolutionary relationships between organisms are subject of phylogeny. The phylogenetic tree showing the evolutionary relationships among various biological species or other entities. All life on earth is part of a single phylogenetic tree, indicating common ancestry. The microbial world has three main cell lineages which are thought to have evolved from a single progenitor. The lineages are formally knowm as Domain. The three domain system, proposed by woese and others, is an evolutionary model of phylogeny based on differences in the sequences of nucleotides in the cell's rRNA, as well as cell's membrane lipid structure and its sensitivity to antibiotics. The phylogenetic tree consist of three domains of organisms the Bacteria and the Archae, cells of which are prokaryotic, and the Eukarya (eukaryotes). Eukaryota and Archaea are more closely related to each other than Bacteria (based on Caliver-Smith's theory of bacterial evolution). Each of these three domains contains unique rRNA. rRNA is the RNA component of the ribosome, which are essential for protein synthesis in all living organisms.

| Property | Archaea | Bacteria | Eukarya |
|---|---|---|---|
| Cell membrane | Glycerol diether linked lipids. | Ester-linked phospholipids and hopanoids. | Ester-linked phospholipids and sterols. |
| Cell wall | Glycoproteins | Peptidoglycan | Various structure |
| Endoplasmic reticulum | Absent | Absent | Present |
| Golgi apparatus | Absent | Absent | Present |
| Lysosome | Absent | Absent | Present |
| Mitochondria | Absent | Absent | Present |
| Chloroplast | Absent | Absent | Present |
| Nucleolus | Absent | Absent | Present |
| RNA polymerase | Many | One | Many |
| Toxin | Sensitive to diphtheria toxin | Resistant to diphtheria toxin | Sensitive to diphtheria toxin |
| Histone | Absent | Present | Present |

**References**

1. Farmer, Jack D. "Role of Geobiology in the Astrobiological Exploration of the Solar System." The Web of Geological Sciences: Advances, Impacts, and Interactions, 2013, doi:10.1130/2013.2500(18).

# Chapter 3
# Eukaryotic Cell Structure and Functions

Eukaryotic cells are cells that contain a membrane-bound nucleus. Eukaryotic cells contain membrane bound organelles, such as endoplasmic reticulum, golgi apparatus, chloroplast, mitochondria and others. The eukaryotic cells are large and observe under the microscope.

Table 1: Eukaryotic cell organelles and their functions

| Organelle | Location | Functions |
|---|---|---|
| **Nucleus** | In animal cell, the nucleus is located in the central region of the cell and in a plant cell, the nucleus is located more on the periphery due to the large water filled vacuole in the center of the cell. | i. Nucleus controls the all metabolic activities of the cell and participates in cell division.<br>ii. Nucleus controls the gene expression and heredity characteristics of an organism.<br>iii. Nucleus is also regulate the passage of ions and small molecule. |
| **Mitochondria** | Mitochondria are the power house of the cell located in the cytoplasm. | i. Mitochondria is to perform cellular respiration.<br>ii. Mitochondria contains enzymes, proteins, lipids and involved in ATP production. |
| **Plasma Membrane** | Outside of a cell. | i. Plasma membranes are selectively permeable membrane.<br>ii. The membrane is to facilitate communication and signaling between cells.<br>iii. Membranes are also involved in the transfer of chemical energy from carbohydrates and fats to ATP. |

| | | |
|---|---|---|
| **Golgi Complex** | Golgi complex is located very near the rough endoplasmic reticulum and the nucleus. | i. The main function is cell secretion, glycosylations |
| **Lysosome** | In liver, spleen and renal cells and cells of immune system- macrophages and polymorphonuclear leukocytes. | i. Lysosome contains hydrolytic enzymes, which functions for extracellular and intracellular digestion. ii. Lysosomes is also involved in digestion and help in fertilization. |
| **Peroxisome** | Found in the cytoplasm. | i. Peroxisomes contain enzymes required for the synthesis of plasmalogens. ii. Participates in the synthesis of cholesterol, bile acids and lipids. |
| | Ribosomes are located inside the animal, human and plant cell. | i. Ribosomes are micromachine for making proteins. ii. Ribosomes act as a catalysis. |
| | Located throughout the cell's cytoplasm. | i. Vacuoles are involved in autophagy, supporting biogenesis and degradation of various structures. ii. Vacuoles are also protect cell from certain bacteria |
| **Chloroplast** | Throughout the cytoplasm of the cells of plant leaves and other plants. | i. The main function of chloroplast is to make food by the process of photosynthesis. |

## References

1. Fawcett, D.W.; The Cell : Its Organelles and Inclusions 2nd edn., Saunders, Philadelphia, 1981.

2. Novikoff, A.B. and E. Holtzmann, Cells and Organelles, 2nd edn., Saunders College Publishing, Philadelphia, 1976.

# Chapter 4

# Journey to the Microbial World

I begin our brief journey to the microbial world by considering different types of microscope. In other words, microbiology and microscopy advance hand in hand. The microscope is the basic tool for microbiologist. Microscopy has played an important role in determining the activity of cells. Microscope is an instrument for viewing objects that are too small to be seen easily by naked eye. In 1590, two dutch eye glass makers, Zaccharis Janssen and Hans Janssen place in a tube. The janssen observed that viewed objects in front of the tube appeared greatly enlarged. The biological microscope mainly consist of an objective lens, ocular lens, lens tube, stage ad reflector. Two most common types of microscopes used today in microbiology are: i. the light microscope and ii. the electron microscope.

## 2.1 Principles of light microscope

Visualization of microorganisms require a microscope either a light microscope or electron microscope. Microscopes are instrument that produce an enlarge image of an object. A light source which may be external to the micrcoscope or built into its base, illuminates the specimen. The light microscope can magnify images upto about 1000 times. In light microscope, the lens systems parts are as follows-

1. Condenser – it collects and focuses light on the material.
2. Objective lens- it produces the image and also magnifies it.

So, light from the condenser lens, and then through the specimen where certain wavelengths to produce an image. Finally, the light passes through the eyepieces lens, which can also be changed to alter the magnification and into the eye. Two types of light microscopes used are: i. simple microscope and ii. compound microscope.

**Simple Microscope**

A simple microscope is a microscope that uses only one lens for magnification. Its magnification is, m= (D/f) for normal adjustment. A simple microscope is used to produce an enlarge image of an object.

**Compound Microscope**

Compound microscope consists of a metal stand or a base from which a short pillar supporting the curved arm rises. A compound microscope is an instrument that is used to view magnified images of small objects on glass slide. The light travels upwards through the condenser and aperture where it passes through the contents of the stage. The light then moves up the head of the microscope where it reaches the eyepiece and is again magnified by the ocular lenses (5X-30X).

**Resolution**

The total magnification of a compound microscope is the product of the magnification of its objective and ocular lenses. Resolution is the function of the wavelength of light used and a characteristic of a objective lens known as its numerical aperture ( a measure of light gathering ability). Higher the numerical aperture (NA), higher will be the degree of resolution. Thus the resolving power of a microscope can be defined in terms of the ability to see neighbouring points in the visual field as two distinct entitites. The resolving power of a microscope is limited by the wavelength of the according to the equation:

$$d = \lambda/2 \, NA$$

Where, d is the minimum distance that two points in the specimen, $\lambda$ is the wavelength of light, NA is numerical aperture.

**2.2 Fluroscence Microscope**

British scientist Sir George g. Stokes first described fluroscence in 1852. He observed that the mineral fluorspar emitted red light when it was illuminated by ultraviolet excitation. Stokes noted that fluroscence emission always occurred at a longer wavelength than of the excitation light. Most cellular components are colourless and cannot be distinguished under a microscope. The basic premise of fluroscence microscopy is to stain the components with dyes. Fluroscence dyes also known as flurophores of flurochromes, are molecules that absorb excitation light at a given wavelength (generally UV). The basic principle of fluroscence microscope is the specimen is illuminated with light of a specific wavelength, which is absorbed by the flurophores causing them to emit light of longer wavelengths. The illumination light is seperatedbfrom the much weaker emitted fluroscence through the use of a spectral emission filter. Four main types of light source are used, including xenon arc lamps or mercury vapour lamps with an excitation filter, lasers, and hgh-power LEDs.

The fluroscence microscope allows viewers to observe the location of certain compound called flurochromes or flurophores. Flurochromes absorb invisible, ultraviolet radiation and release a portion of the energy in the longer visible wavelengths, a phenonmenon is called fluroscence. The fluroscence microscope is used to visualize specimens that flurosce, that is emit light of one colour when light of another colour shines upon them. Fluroscence is the emission of light by a substance that has absorbed light or other electromagnetic substances. On 8 october 2014, the nobel prize is chemistry was awarded to Eric Betzig, William Moerner and Stephan Hell for the development of super-resolved fluroscence microscopy. A flurochrome (such as rhodamine) covalently linked to an antibody to produce a fluroscence antibody that can be used to determine the location of a specific protein within the cell, this technique is called immunofluroscence. DAPI (diamidino-2-phenylindole) is a fluroscent dye that binds strongly to adenine-thymine rich regions in DNA. Hoechst stains are part of a family of blue fluorescent dyes used to stain DNA. A major example of fluroscent stain is phalloidin which is used to stain actin fibers in mammalian cells. The GFP (green fluroscent protein) was first isolated from the jellyfish, Aequorea victoria. GFP-tagging is a way of preparing a sample for fluroscence microscopy by using the GFP as a fluroscent protein repoter.

## Application

i. Flurochromes can aso be used to locate DNA and RNA molecules that contain specific nucleotides.

ii. flurochromes have been used to study sizes of molecules.

iii. To identify structers in fixed and live biological samples.

## 2.3 Electron Microscopy

Electron microscopy is a technique for obtaining high resolution images of biological and non biological specimen. Electron microscope uses electrons to illuminate a specimen and create an enlarged image. The first electron microscope prototype was built in 1931 by german engineers Ernst ruska and Max knoll. In an electron microscope, a beam of electrons is projected from an electron gun and passed through a series of electromagnetic lenses. The condenser lens collimates the electron beam on the specimen and an enlarge image is produced by a series of magnifying lenses. Electron microscope have much greater resolution. There are two types of electron microscope as described below:

        i. Transmission electron microscope.

        ii. Scanning electron microscope.

### Transmission Electron Microscope

In 1986, Ruska was awarded the Nobel Prize in physics for the development of transmission electron microscopy. The transmission electron microscope is very powerfull tool for material science. The transmission electron microscope can provide vastly greater resolution than the light microscope. The TEM (transmission electron microscope) operates on the same basic principles as the light microscope but uses electron instead of light. When an electron beam passes through a thin section specimen of a material, electrons are scattered. The limit of resolution of a microscope is directly proportional to the wavelength of the illuminating light. The wavelength of an electron varies with the speed at which the particle is travelling, which in turn depnds on the accellarting voltage applied in the microscope.

## Application

i: It is helpfull to study the internal structure of cells.

ii: The ultrastructures of different cellular organelles can be studied by TEM.

iii: The packaging of chromosome into nucleosome and its higher level can be revealed by TEM.

iv: TEM helps in the study of antigen antibody reaction, study of malignant cells etc.

v: TEMs provide topographical, morphological, compositional and crystalline information.

vi: Colleges and Universities can utilize TEMs for research and studies.

## Scanning Electron Microscope

The scanning electron microscope (SEM) is a powerful magnification tool that utilizes focused beams on electron to obtain information. The high resolution, three dimensional images produced by the SEM. The basic principle is that a beam of electrons is generated by a suitable source, typically a tungsten filament or a filled emission gun. The electron beam is accerlated through a high voltage (e.g 20KV) and pass through a system of apertures and electromagnetic lenses to produce a thin beam of electrons. Then the beam scans the surface of the specimen. Electrons are emitted from the specimen by the action beam and collected by a suitably positioned dectector. The microscope operator are watching the images on a screen. The scanning electron microscope allows visualization of surface feature of a solid sample by scanning through an electron beam. In advance SEM machine magnification can range from 10X to 5000X and resolution of about 1 nm. The main SEM components include:

- Source of electrons.
- Column down which electrons travel with electromagnetic lenses.
- Electron detector.
- Sample chamber.
- Computer and display to view the images.

## Application

i: It is usefull in studies of eternal structures of bacterial cells, spores and fungi.

ii: It helps to study the morphological changes in tissues infect with microorganisms.

iii: SEMs are used in materials science for research, quality control and failure analysis.

## Differences between SEM and TEM

| Transmission Electron Microscope | Sanning Electron Microscope |
|---|---|
| Electron beam passes through thin sample. | Electron beam scan over surface of sample. |
| Specially prepared thin samples are supported on TEM grids. | Sample can be any thickness and is mounted on aluminium stab. |
| Image shown on fluroscent screen. | Image shown on TV monitor. |
| Specimen stage halfway down column. | Specimen stage in the chamber at the bottom of the column |
| TEM is based on transmitted electron. | SEM is based on scattered electrons. |

## Differences between Light microscope and Electron microscope

| Light microscope | Electron Microscope |
|---|---|
| Can view both live and dead specimens. | Views only dead specimens. |
| Uses light rays to illuminate specimen. | Uses beam of electrons to view specimens. |
| Lenses are made up of glass. | Lenses are made of electromagnets. |
| Low magnification up to 1500X | High magnification of 100,000X to 300,000X. |
| Image is coloured. | Image is black and white. |

# Chapter 5

# Bacteria

**B**acteriology is the science that deals with the study of organisms known as bacteria (singular, bacterium). The word germ is probably synnomous with bacterium. Bacteria among the simplest form of life know and hence show characteristic of both plants and animals. Bacteria are small (microscopic size) organisms that can be found in most environments, for example in soil, water, and on and inside the human body. There are around 50 million bacteria in every gram of surface soil. Bacteria can be harmful, but some species of bacteria are needed to be healthy. Research suggests that efforts to make a cleaner environment, free from bacteria, are contributing to the rise in obesity, cancer and heart disease.

**Size**

Bacteria are considerably smaller than yeasts, moulds, algae, and protozoa. They vary greatly in size according to species. Bacteria show considerable variation in size according to species. Regardless of their size none are visible with naked eye. Some bacteria are large enough to see with naked eye. Whereas Schaudinnum butschlii that measure between 4 and 5 µm in diameter are considered to large bacteria.

**Table 3.1 Size of bacteria**

| Bacteria | Size (µm) |
|---|---|
| Cocci | 0.5 µm to 1.25 µm |
| Bacillus or rod shaped bacteria | 0.5 µm -1.0 µm X 2-3 µm |
| Helical or spiral bacteria | 1.5 µm |
| Euplopiscium fishelsohni | 200 µm X 80 µm |
| Thiomargarita namibiensis | 750 µm |

**Morphology (shape) of bacterial cell**

The three basic bacterial shapes are coccus (spherical cell), bacillus (rod shaped), and spiral (twisted forms). Cocci are bacteria that are spherical, ovoid, or generally round in shape. Cocci occur in single cell or remain attached, that can be based on cellular arrangement. Cocci are of different types:

- Diplococci = Occur in pairs. e.g Streptococcus pneumonia and Neisseria gonorrhoeae, Moraxella catarrhalis.

- Styphalococci = Irregular (graph lke) cluster of cocci. e.g Styphalococcus aureus.

- Streptococci = Occur in long chain. e.g Streptococcus lactis, Streptococcus pyogene.

- Tetracocci = Tetrads are square arrangement of four cocci. e.g Aerococcus, Pediococcus and Tetragenococcus.

- Sarcinae = The cocci are arranged in a cuboidal manner made up of 8 or more cells. e.g Sarcinae ventriculi, Sarcinae ureae etc.

### Arrangement of baccilus

The cylindrical or rod shaped bacteria are called bacillus. They may be motile or non motile.

- Diplobacilli = Occur in pair. e.g bacillus cereus.

- Streptobacilli = Occur in chain. e.g Bacillus moniliformis.

### Spiral bacteria

Spiral are curved bacteria which can range from a gently curved shaped. Many spirili are rigid and capable of movement.

Besides the above groups, following other shapes of bacteria are also present –

- Vibrio = They are comma shaped bacteria with less than one complete turn or twist in the cell. e.g Vibrio chlorea.

- Filamentous = Some bacteria are filament like. e.g Beggiatoa, Thiothrix.

- Spirilla = They have a rigid spiral structure. e.g Helicobacter pylori, Spirillum winogradskyi.

- Pleomorphic = Some bacteria are able to change their shape and size in response to variation in the surrounding environment. In pure culture, they can be observed to different shapes.

### Structure of bacterial cell wall

The cell wall is usually fairly rigid just outside the plasma membrane. A bacterial cells show typical prokaryotic structure. Species of bacteria can be divided into two major groups, called gram positive and gram negative. The distinction between gram - positive and gram - negative based on the gram stain reaction. Gram positive bacteria stained purple whereas gram negative bacteria were coloured, pink or red. Bacterial cell wall composed of: i. Peptidoglycan. ii. Outer membrane and iii. Surface membrane.

## Peptidoglycan

Peptidoglycan is a polymer composed of two sugar derivatives – N-acetylglucosamine and N-acetylmuramic acid and a few amino acids, including L-alanine, D-alanine, D-glutamic acid and either lysine or diaminopimelic acid (DAP). The sugar component consists of alternating residues of beta (1,4) linked N = acetylglucosamine (NAG) and N – acetylmuramic acid. The peptidoglycan layer is thicker in Gram positive bacteria (20-80 nanometers) then in Gram negative bacteria (7-8 nanometers). The gram positive bacteria have substances called teichoic and teichuronic acids interspersed with the peptidoglycan polymer.

**Table 3.1 Comparision of the cell walls of Gram-positive and Gram-negative bacteria**

| Gram-positive cell wall | Gram-negative cell wall |
| --- | --- |
| 1. Thick peptidoglycan layer. | 1. Thin peptidoglycan layer. |
| 2. Principal surface antigen is teichoic acid. | 2. Principal surface antigen is lipopolysaccharides. |
| 3. Cell wall is rigid. | 3. Cell wall is elastic. |
| 4. Cell wall is single layered. | 4. Cell wall is multilayered. |
| 5. Teichuronic acid is present. | 5. teichuronic acid is absent. |
| 6. Presence of glycolipids. | 6. Absence of glycolipids. |
| 7. Some examples of gram-positve bacteria are Streptococcus, Clostridium, Lactobacillus etc | 7. Some examples of gram negative bacteria are Vibrio, Rhizobium, Escherichia coli, Acetobacter. |

## Gram staining method

This technique was developed by Danish physician, Hans Christian Gram, in 1884; which is very useful in taxonomic grouping of bacteria.

### Required reagents:

- Gram's Iodine.
- acetone- alcohol
- Safranin

### Procedure

1. Prepare the smear from the specimen.

2. Allow the smear to air-dry completely.

3. Rapidly pass the slide three times through the flame.

4. Cover the fixed smear with crystal violet for 1 minute and wash with distilled water.

5. Tip off the water and cover the smear with gram's iodine for 1 minute.

6. Wash of the iodine with clean water.

7. Decolorize rapidly with acetone-alcohol for 30 seconds.

8. Wash off the acetone-alcohol with clean water.

9. Cover the smear with safranin for 1 minute.

10. Wash off the stain wipe the back of the slide. Let the smear to air-dry.

11. Examine the smear with oil immersion objective to look for bacteria.

**Result**

Gram positive bacterium..............................Purple colour

Gram negative bacterium..............................Pink colour

### Lipopolysaccharide

Lipopolysaccharide is localized in the outer layer of membrane and is, in noncapsulated strains, exposed on the cell surface. Bacterial lipopolysaccharides (LPS) are the major outer surface membranes components present almost all gram negative bacteria. It comprises three parts: i. O antigen (or O polysaccharides) ii. Core oligosaccharide, and iii. Lipid A.

O antigen: The O antigen is repeating oligosaccharide unit typically comprised of two to six sugars. The core domain attached to lipid A and commonly contain sugars such as heptose and 3-deoxy-D-manno-2-octo-ulosonic acid (known as KDO, keto-deoxyoctulosonate). LPS contributes to the negative charge on the bacterial surface.

### Components external to the cell surface

Bacterial cells possess various structures external to the cell wall that basically contribute in protection, attachment to objects, and cell movement. Several of these discussed

### The Glycocalyx : Slime Layer and Capsule

The glycocalyx represents the gelatinous covering around many bacterial cells secreted at the time of their active growth. When the glycocalyx does not from a persistent layer, but it present more diffusedly forming a loose mass around the bacterial cell, it is called slime layer. Capsules or slime layer may be thick or thin and rigid or flexible, depending on their chemistry and degree of hydration. Capsules are composed generally of polysaccharides (e.g Streptococcus mutans, S. salivarious, Xanthomonas corynebacterium) that contain, apart from glucose, amino sugars, rhamnose, uronic acids of various sugars, 2-keto-3-deoxygalactonic acid, and organic acids such as pyruvic acid and acetic acid. Capsules can be seen negative staining with dyes. Capsules protect bacterial cell against phagocytosis by various protozoans and WBCs. Capsules protect bacterial cells from dessiccation. Capsules prevent bacteriophages and detergents from attaching on to bacterial surfaces.

**Pili and Fimbriae**

Many prokaryotes have short, fine, hair like appendages that are thinner than flagella. They are usally called fimbriae. Fimbriae are responsible for more than attachment. Fimbriae are major factor for bacterial virulence. Each fimbriae are special protein called adhesions. Pili are similar to fimbriae but typically longer structure.

**The Flagella**

The structure of bacterial flagella have been described by Simon (1978), Doestsch and Sjoblad (1980) and Ferris and Beveridge (1985). The most common means of location in prokaryotes is via the rotation of flagella. Bacterial flagella are slender, rigid structures, about 20 nm across and up to 15 or 20 µm long. Flagella are so thin they cannot be observed directly with bright field microscope. A single cell may have many flagella spread all over the surface of the cell (peritrichous flagella), the flagella may be polar, found at one (monotrichous if single) or both ends (amphitrichous if only two in total). Lophotrichous bacteria have a cluster of flagella at one or both ends. Transmission electron microscope shows the bacterium flagellum consists of three parts: i. Filament. ii. Hook. iii. Basal body.

The filament is hollow, rigid slender constructed of subunit of the protein flagellin, which ranges in molecular weight from 30,000 to 60,000 daltons. Basal body consists of the L,P,C and MS rings.

L-ring: L-ring is the outer ring present only in gram negative bacteria, it anchored lipopolysaccharide layer.

P-ring: P-ring anchored in peptidoglycan layer of cell wall.

MS ring anchored in cytoplasmic membrane and C ring anchored in cytoplasm.

Hook: The wider region at the base of the flagellum called hook. It connects filament to the motor protein in the base.

Many bacteria have a gliding motility, which allows flagella-free bacteria to move rapidly across surfaces. A very different type of motility, gliding motility in bacteria: Cyanobacteria, Cyanophages, and some mycoplasms. Many bacterial cells are motile by means by flagella.

# Procedure of flagella staining

The flagella stain allows observation of bacterium flagellum under the microscope. Detail flowchart of the flagella staining procedure described below:

Grow the organisms to be stained at room temperature on blood agar for 16 to 24 hours.

Add a small drop of water to a microscope slide.

Dip a sterile inocuting loop into sterile water.

Touch the loopful of water to the colony margin briefly.

Touch the loopful of motile cells to the drop of water on the slide and cover the slide with cover slip.

Examine the slide immediately under 40x for motile cells. if motile cells are seen leave the slide for 5-10 minute.

Then, apply two drops of RYU flagella stain gently on the edge of the cover slip. The stain will flow by capillary action and mix with the cell suspension.

↓

After 5-10 minutes at room temperature, examine the cells for flagella.

↓

Cells with flagella may be observed at 100x.

## The Bacterial Endospore

Certain species of bacteria produce a special resistant, dormant structure called endospore during a process called sporulation. Endospore developed within vegetative bacterial cells of several genera: Bacillus and Clostridium (rod), Sporosarcina (cocci) and others. These structures are resistant to environmental stresses such as heat, UV, gamma radiation etc.

Structure: Endospore consists of a central protoplast and the core. The core is composed of DNA, t-RNA, enzyme, ribosome etc. The core is covered by a thin membrane called core membrane or inner membrane or germ cell membrane.

## Endospore Formation

The process of forming an endospore is complex. The model organism used to study endospore formation is Bacillus subtilis. Endospore development requires several hours to complete. Key morphological changes in the process have been used as markers to define stages of development. As a cell begins the process of forming an endospore, it divides asymmetrically (Stage II). This results in the creation of two compartments, the larger mother cell and the smaller forespore. These two cells have different developmental fates. Intercellular communication systems coordinate cell-specific gene expression through the sequential activation of specialized sigma factors in each of the cells. Next (Stage III), the peptidoglycan in the septum is degraded and the forespore is engulfed by the mother cell, forming a cell within a cell. The activities of the mother cell and forespore lead to the synthesis of the endospore-specific compounds, formation of the cortex and deposition of the coat (Stages IV+V). This is followed by the final dehydration and maturation of the endospore (Stages VI+VII). Finally, the mother cell is destroyed in a programmed cell death, and the endospore is released into the environment. The endospore will remain dormant until it senses the return of more favorable conditions.

## Endospore Staining

In 1922, Dorner published a method for staining endospore. Shaeffer and Fulton modified Dorner's method in 1933 to make the process faster.

Principle

Bacterial endospore are metabolically inactive, highly resistant structures produced by some bacteria as a defensive strategy against unfavourable environmental condition.

In the Schaeffer-Fulton's method, primary stain malachite green is forced into the spore by steaming the bacterial emulsion. Malachite green is a water soluble and has a low affinity for cellular material. So, vegetative cells may be decolorized with water. Safranin is then applied to counterstain any cells which have been decolorized. At the end of the staining process, vegetative cells will be pink and endospore will be dark green.

Reagents used for endospore staining

Primary stain: Malachite green 0.5% (wt/vol) aqueous sol$^n$, 0.5 gm of malachite green and 100 ml of distilled water.

Decolorizing agent: Tap water or distilled water.

Counter stain: Safranin.

Procedure of endospore staining:

1. Take a clean grease free slide and make smear.

2. Air dry and heat fix the organism on a glass slide and cover with a square of blotting paper or toweling cut to fit the slide.
3. Saturate the blotting paper with malachite green stain solution and steam for 5 minutes, keeping the paper moist and adding more dye as required. Alternatively, the slide may be steamed over a container of boiling water.
4. Wash the slide in tap water.
5. Counterstain with 0.5% safranin for 30 seconds. Wash with tap water; blot dry.
6. Examine the slide under microscope for the presence of endospores. Endospores are bright green and vegetative cells are brownish red to pink.

# Chapter 6

# Microbial Genetics of Bacteria

For all living organisms, gene expression is always happens. Expression of genes is triggered by extracellular signals. The word 'gene' refers to functional unit of DNA that can be transcribed. Expression of genes is triggered by extracellular stimuli. Cells transform external stimuli. Of all living organisms, genes expression highly regulated in both prokaryotes and eukaryotes. External stimuli (such as hormones and neurotransmitter) cells to change their behaviour. Signal transduction pathways are always activated a chemical signaling. The expression of gees are regulates DNA-binding proteins.

Bacteria have long played key role in genetics. To map genes in bacteria, geneticists use essentially the same experimental strategies. Crosses are made between strains that differ in genetic markers, and recombinants (exchange of genetic material) are detected and counted. Bacterial genomes sequenced include 4.6-million-base-pair (4.6-megabase-pair (Mb) genome of E.coli, the 1.44 Mb genome of the Lyme disease causative agent Borella burgdorferi. The genetic material in bacteria is stored in two structures-a single main chromosome, carrying a few thousand genes. Bacterial genetics is the study of how genetic information is transferred, either from a particular bacterium to its offspring. DNA is eventually the genetic material of all living organisms, from bacteria to humans. So, genetic information in bacteria is encoded in DNA. In bacteria, genetic information takes place through interchange of genetic material. Genetic material can be transferred between bacteria bt three main processes:

1. Conjugation.

2. Transformation.

3. Transduction.

**Conjugation**

Conjugation is a process which promotes DNA transfer from a donor to a recipient cell mediated by physical contact. The process was first discovered by Joshua Lederberg and Edward Tatum (1946) in Escherichia coli. Plasmid is transferred to a recipient cell from a donor.

**Mechanism:** Donor cells produces pilus (pilus is hair like appendage required for bacterial conjugation). The transfer of genetic material is always unidirectional, with male chromosome moving into the potential female. The converse movement of the female genes into male cells never occurs. There are two mating types of bacteria, one is male type or F⁺ or donor cell and other one is F⁻ or recipient cell. F⁺ or donor cells are capable of transferring genes into female cells. the F-factor is a closed circular double-stranded DNA molecule present in one copy per cell. When F⁺ cells are mixed with F⁻ cells, conjugal pairs are formed by the attachment of a male(F⁺) sex pilus to the surface of a female cell. The transfer of F-factor from an F⁺ or Hfr cells to and F⁻ cells takes place in 90-100 minutes.

## High frequency recombination

There is a another type of conjugation where passage of nucleoid DNA takes place through conjugation tube. Strains of bacteria are known as Hfr (High frequency recombination) strain. The Hfr cells from a sex pilus and attaches to a recipient F⁻ cell. DNA begins to be transferred from the Hfr cell to the recipient cell.

## Transformation

Transformation is discovered by Fred Griffith in 1928. Genes can be transferred without cellular contact or vectors. Transformation is a genetic transfer process by which free DNA or foreign DNA is incorporated into a recipient cell. Plasmids can be used as vectors to carry foreign DNA into a cell. Frederick Griffith (1928) a medical officer in the British Ministry of Health conducted the famous experiment on mice to explain the chemical basis of heredity. He selected two strains of the celebrated bacteria known to cause pneumonia (Streptococcus pneumonia). One type has rough (R) non-capsulated cells and another one with smooth (S) capsulated cells. The R type is non-pathogenic while the S type is pathogenic. The process of transformation is mentioned below:

1. When R type non pathogenic and S type dead pathogenic cells are injected in mice, the mice remain alive respectively.

2. When pathogenic (S type) cells are injected in mice, they suffer from pneumonia and died.

3. When live non-pathogenic (R-type) cells are mixed with dead pathogenic (S-type) cells are injected in mice, they also suffered from pneumonia and died.

After 16 years, Oswald Avery, C.M Mecleod and M.J. Mccarty in 1944 published their research article and provided the first evidence that Griffith's transforming principle is DNA, and hence DNA is the carrier of genetic information.

# Transduction

The transfer of genetic material from one cell to another by a bacteriophage is called transduction. Viruses that infect bacteria are called bacteriophages or phages. This process was first discovered by Norman Zinder and J. Lederberg in 1952 during their experiments to see whether the process of conjugation existed in salmonella. Transduction is a common tool used by scientist to introduce different DNA sequences of interest into a bacterial cell.

## Types of transduction

There are two types of transduction:

1. Generalized transduction.

2. Specilized transduction.

Generalized transduction: In generalized transduction, a bacterial host cell infected with either virulent or temperate bacteriophage engaging in the lytic cycle of replication. It is characterized by random nature of DNA fragments and genes transduced. They inject bacterial DNA from one bacterium to another and not the viral DNA. The generalized transduction is said to be complete when the transduced DNA get integrated with the bacterial chromosome and also replicated and passes on to all daughter cells. This ability of the bacteriophage to carry with its genetic material any region of the bacterial DNA is now known as generalized transduction.

Specilized transduction: Specialized transduction involves the transduction of bacterial and viral DNA. It is mediated by temperate bacteriophages whose chromosomes are able to integrate at one or a few specified attachment sites on the host chromosome. The bacteriophage randomly attached to its host cell, injecting viral DNA inside.

## References

1: Griffiths AJF, Miller JH, Suzuki DT, et al. Bacterial Conjugation. In An Introduction to Genetic Analysis: 7th edition (2000).

2: Meibom, Karin L.; et al. "Chitin induces natural competence in Vibrio cholerae." Science 310.5755 (2005): 1824-1827. PubMed PMID: 16357262.

3: Achtman, M. 1975. Mating aggregates in Escherichia coli conjugation. J. Bacteriol. 123:505–515.

4: Achtman, M., N. Kennedy, and R. Skurray. 1977. Cell-cell interactions in conjugating Escherichia coli: role of traT protein in surface exclusion. Proc. Natl. Acad. Sci. USA 74:5104–5108.

5: Achtman, M., P. A. Manning, C. Edelbluth, and P. Herrlich. 1979. Export without proteolytic processing of inner and outer membrane proteins encoded by F sex factor tra cistrons in Escherichia coli minicells. Proc. Natl. Acad. Sci. USA 76:4837–4841.

6. Holmes RK, Jobling MG. Genetics. In: Baron S, editor. Medical Microbiology. 4th edition. Galveston (TX): University of Texas Medical Branch at Galveston; 1996. Chapter 5.

7. Tortora, Gerard J., Berdell R. Funke, Christine L. Case. Microbiology: An Introduction. Redwood City: CA: Benjamin/Cummings Publishing Company, Inc., 2001.

8. Abdel-Monem, M., H. Durwald, and H. Hoffmann-Berling. 1976. Enzymic unwinding of DNA. II. Chain separation by an ATP-dependent DNA unwinding enzyme. Eur. J. Biochem. 65:441–449.

9. Abdel-Monem, M., G. Taucher-Scholz, and M. Q. Klinkert. 1983. Identification of Escherichia coli DNA helicase I as the traI gene product of the F sex factor. Proc. Natl. Acad. Sci. USA 80:46594663.

10. Abo, T., S. Inamoto, and E. Ohtsubo. 1991. Specific DNA binding of the TraM protein to the oriT region of plasmid R100. J. Bacteriol. 173:6347–6354.

11. Abo, T., and E. Ohtsubo. 1993. Repression of the traM gene of plasmid R100 by its own product and integration host factor at one of the two promoters. J. Bacteriol. 175:4466–4474.

12. Hyg (Lond). 1928 Jan;27(2):113-59. The Significance of Pneumococcal Types. Griffith F.

13. J Exp Med. 1944 Feb 1;79(2):137-58. Studies on the chemical nature of the substance inducing transformation of Pneumococcal types: induction of transformation by a desoxyribonucleic acid fraction isolated from Pneumococcus type III. Avery OT, Macleod CM and McCarty M.

14. Baltrus, David A., and Karen Guillemin. "Multiple phases of competence occur during the Helicobacter pylori growth cycle." FEMS microbiology letters 255.1 (2006): 148-155. PubMed PMID: 16436074.

15. Achtman, M., P. A. Manning, B. Kusecek, S. Schwuchow, and N. Willetts. 1980. A genetic analysis of F sex factor cistrons needed for surface exclusion in Escherichia coli. J. Mol. Biol. 138:779–795.

16. Achtman, M., G. Morelli, and S. Schwuchow. 1978. Cell-cell interactions in conjugating Escherichia coli: role of F pili and fate of mating aggregates. J. Bacteriol. 135:1053–1061.

17. Achtman, M., and R. Skurray. 1977. A redefinition of the mating phenomenon in bacteria, p. 234–279. In J. L. Reissig (ed.), Microbial Interactions: Receptors and Recognition, ser. B, vol. 3. Chapman & Hall, Ltd., London

# Chapter 7

# Overview of Viruses

In 1900, it was generally accepted that many of the recognized human diseases were caused by microorganisms, the first evidence of viruses as causative agents of disease came in 1892 when Ivanowski showed that cell free extracts of diseased tobacco leaves passed through bacteria proof filters, could cause disease in healthy plants. The word virus come from a latin word simply meaning slimy fluid. The first human virus described was the agent which causes yellow fever. The virus was discovered and reported in 1901 by the US Army physician Walter Reed. Beijerinck (1897) coined the Latin name virus meaning poison. Viruses are very small- smaller than the smallest cell. The scientific study of viruses are called virology. Viruses were first discovered after the development of a porcelain filter called the chamber-land-pasteur filter, which could remove all bacteria visible in the microscope from any liquid sample. Thus virology (study of viruses) is a significant part of microbiology. Viruses cause many more illnesses disease.

## General properties of viruses

The virus is made up of a genetic information molecule and a protein layer that protects that information molecule. The core of the virus is made up of nucleic acids, which then make up the genetic information in the form of DNA or RNA. Viruses are much smaller than prokaryotic or eukaryotic cells. Some viruses (bacteriophage) are infect prokaryotic cell or eukaryotic cells. Viruses are non cellular, self replicating agents. They are transmitted very easily from one organism to another organism.

## The Structure of Viruses

Viruses come in amazing variety of size and shapes. They are very small and are measured in nanometers, which is one-billionth of a meter. Viruses are seen with a scanning electron microscope. A simple virus particle often called a virion. Virions range in size from about 10 to 400 nm in diameter. Viruses contain either DNA or RNA but not both. Each virion contains only one molecule of nucleic acid, called genome. The protein coat surrounding the genome is called capsid. The capsid is made up of large number of protein subunit called capsomeres.

**Symmetry:** The capsid is symmetrically arranged around the central nucleic acid. Capsids protect genome from atmosphere. Viral capsids are made of repeated protein subunits. There are three types of capsids symmetry:

    i. Cubical (icosahedral)

    ii. Helical Capsids.

    iii. Complex capsids.

i: Cubical (icosahedral) capsids: These viruses appear spherical in shape. The icosahedral made up of equilateral triangles. They have a polygon with 20 sides (facets) and 30 edges. They usally made up of five or six pentamers. Icosahedral capsids contain both pentamers and hexamers. Simian virus 40 (SV 40) only pentamers.

ii: Helical capsids: Helical capsids are shaped like hollow tube or central cavity that is made by proteins arranged in a circular fashion, creating a disc like shape. They are usally 15-19 nm wide. e.g Tobacco Mosaic virus (TMV), Influenza virus etc

iii: Complex capsids: These virus structures have a combination of icosahedral and helical shape. The head of the virus has an icosahedral shaped within a helical shaped tail. e.g Pox virus and bacteriophages like $T_2, T_4$ and $T_6$.

## Viriod

Theoder O. Diener discovered a cell-invading plant pathogen 80 times smaller than a viruses; the viriod. Viriods are unique plant pathogens consisting of low molecular weight, autonomously replicated single stranded RNA molecules (approximately 246 to 425 nucleotides (nt) without any functional open reading frame (ORF) in their genome. The first viriod, potato spindle tuber viriod (PSTVd), was discovered in the late 1960s – early 1970s (Diener, 1971). Viriods are classified into two families : the pospiviroidae, which replicate in the nucleus, and the Avsunviroidae, which replicate in the chloroplast (Tsagris, 2008). They do not code for protein. The viriods potato spindle tuber viriod (PSTVd) and Citrus exocortis viriod (CEVd) are both replicated in the nucleus.

## Prion

The word itself derives from proteinaceous infectious particle meanning that the infectious agent consists only of protein with no nucleic acid genome. The term prion refers to abnormal pathogenic that are transmissible and are able to induce abnormal folding of specific normal cellular proteins called prion proteins. Prions cause a variety of neurogenerative diseases in humans and animals. Prions are responsible for bovine spongiform encephalopathy (BSE or mad cow disease), Scarpie (goat and sheep), chronic wasting disease (deer and elk), and the human diseases kuru, fatal, sporadic insomnia, Gerstmann-Strausster-Scheinzer disease, Creutzfeldt-Jacob disease (CJD). In 1950s, disease known as kuru, in Papua New Guinea discovered; neuropathological similarity between kuru, CJD, and serapie noted. Research suggests that scrapie is caused by abnormal form of a cellular protein. The abnormal form is called PrP$^{sc}$ (for scrapie-associated prion protein) and the normal cellular form is called PrP$^c$.

## Swine flu

Swine flu is also known as swine influenza, hog flu and pig flu. Swine flu is an infection caused by a virus. Swine flu can spread from human to human through droplets and aerosols created while coughing and sneezing by an infected person. Swine flu is a disease of pigs that can be passed to humans. In 2017, swine flu was normally of the H3N2 influenza subtype. Swine influenza is an acute respiratory disease caused by influenza virus.

### Symptoms:

Swine flu has an incubation period (the time taken for the symptoms to appear) of 1-4 days. Its symptoms are include:

- Fever.
- Cough.
- Headache.
- Muscle and joint pain.
- Sorethroat and runny nose.
- Vomiting and diarrhoea.
- Body aches.
- Chills.
- Fatigue or tiredness, which can be extreme.

### Prevention:

You can help prevent the spread of germs that cause respiratory illnesses like influenza by:

- Wash your hand often with soap and water.
- Cover your mouth and nose with a tissue paper when you cough or sneeze.
- Avoid touching your eyes, nose or mouth.
- Staying home from work or school if you are sick.

## AIDS (Acquired immunodeficiency virus)

Acquired immunodeficiency syndrome (AIDS) is a disease of human immune system caused by human immunodeficiency virus (HIV). The human immunodeficiency virus (HIV) is an RNA virus known as a retrovirus. HIV is the member of the lentivirus family. In 1981, it has killed more than 25 million people. HIV cannot be transmitted through causal, everyday contact. Mosquito and other insects do not transmit HIV. HIV is primarilly spread through unprotected vaginal or anal intercourse with someone who is HIV positive, by sharing contaminated needles, syringes and other injecting equipment.

# References

1. Griffiths AJF, Miller JH, Suzuki DT, et al. Bacterial Conjugation. In An Introduction to Genetic Analysis.7th edition (2000).

2. Meibom, Karin L., et al. "Chitin induces natural competence in Vibrio cholerae." Science 310.5755 (2005): 1824-1827. PubMed PMID: 16357262.

3. Achtman, M. 1975. Mating aggregates in Escherichia coli conjugation. J. Bacteriol. 123:505–515.

4. Achtman, M., N. Kennedy, and R. Skurray. 1977. Cell-cell interactions in conjugating Escherichia coli: role of traT protein in surface exclusion. Proc. Natl. Acad. Sci. USA 74:5104–5108.

5. Achtman, M., P. A. Manning, C. Edelbluth, and P. Herrlich. 1979. Export without proteolytic processing of inner and outer membrane proteins encoded by F sex factor tracistrons in Escherichia coli minicells. Proc. Natl. Acad. Sci. USA 76:4837–4841.

6. Holmes RK, Jobling MG. Genetics. In: Baron S, editor. Medical Microbiology.4th edition. Galveston (TX): University of Texas Medical Branch at Galveston; 1996. Chapter 5.

7. Tortora, Gerard J., Berdell R. Funke, Christine L. Case. Microbiology: An Introduction. Redwood City: CA: Benjamin/Cummings Publishing Company, Inc., 2001.

8. Abdel-Monem, M., H. Durwald, and H. Hoffmann-Berling. 1976. Enzymic unwinding of DNA. II. Chain separation by an ATP-dependent DNA unwinding enzyme. Eur. J. Biochem. 65:441–449.

9. Abdel-Monem, M., G. Taucher-Scholz, and M. Q. Klinkert. 1983. Identification of Escherichia coli DNA helicase I as the traI gene product of the F sex factor. Proc. Natl. Acad. Sci. USA 80:46594663.

10. Abo, T., S. Inamoto, and E. Ohtsubo. 1991. Specific DNA binding of the TraM protein to the oriT region of plasmid R100. J. Bacteriol. 173:6347–6354.

11. Abo, T., and E. Ohtsubo. 1993. Repression of the traM gene of plasmid R100 by its own product and integration host factor at one of the two promoters. J. Bacteriol. 175:4466–4474.

12. Hyg (Lond). 1928 Jan;27(2):113-59. The Significance of Pneumococcal Types.Griffith F.

13. J Exp Med. 1944 Feb 1;79(2):137-58. Studies on the chemical nature of the substance inducing transformation of Pneumococcal types: induction of transformation by a desoxyribonucleic acid fraction isolated from Pneumococcus type III. Avery OT, Macleod CM and McCarty M.

14. Baltrus, David A., and Karen Guillemin. "Multiple phases of competence occur during the Helicobacter pylori growth cycle." FEMS microbiology letters 255.1 (2006): 148-155. PubMed PMID: 16436074.

15. Achtman, M., P. A. Manning, B. Kusecek, S. Schwuchow, and N. Willetts. 1980. A genetic analysis of F sex factor cistrons needed for surface exclusion in Escherichia coli. J. Mol. Biol. 138:779–795.

16. Achtman, M., G. Morelli, and S. Schwuchow. 1978. Cell-cell interactions in conjugating Escherichia coli: role of F pili and fate of mating aggregates. J. Bacteriol. 135:1053–1061.

17. Achtman, M., and R. Skurray. 1977. A redefinition of the mating phenomenon in bacteria, p. 234–279. In J. L. Reissig (ed.), Microbial Interactions: Receptors and Recognition, ser. B, vol. 3. Chapman & Hall, Ltd., London

# Chapter 8
# Mycoplasma

Mycoplasma belongs to the class molicutes. Mycoplasmas are the smallest organisms found in the respiratory and genital tracts of man and many animal species. They are also found in plants, insects, soil and sewage. Mycoplasma (Greek: mykes, fungus; plasma, something moulded) refers to the filamentous nature of organisms of some species. In 1898, Nocard and Roux were to first to isolate a mycoplasma species in culture from bovine. So, mycoplasmas are distributed widely in nature.

Mycoplasmas grew in cell free bacteriological media. They lack a rigid cell wall containing peptidoglycan. Mycoplasma are self replicating bacteria. More than 120 different species of mycoplasma and 7 species of ureplasma are found. Mycoplasma require sterol for growth while ureplasma species require urea for fermentation. Mycoplasma is also called as PPLO (Pleuro pneumonia like organisms). Mycoplasmas are relatively resistant to pencillins, cephalosporins, and sensitive to tetracyclnes, and several other antibiotics. Some species of Mycoplasmas are M.adleri, M.agalactiae, M.anatis, M.bovis. Mycoplasmas disease is transmitted by droplets infections of nasopharyngeal secretions. Several species of mycoplasmas e.g. M.pneumonia infection is common in school aged children. They are composed of minute cells. Mycoplasmas are spherical to filamentous cell with the smallest genomes (a total of about 500 to 1000 genes); they are low in guainine and cytosine. Mycoplasmas have been nicknamed crabgrass of cell cultures because their infections are persistent.

# Chapter 9
# Genetic Engineering

Genetic engineering refers to the direct manipulation of DNA to alter an organism's characterstics in particular way. The basic technique of the genetic engineering is cut the DNA of interest by restriction enzyme and this fragments of DNA is manipulated.

Restriction enzyme is also called as molecular scissors. They are widespread both bacteria and archaea but very rare in eukaryotes. Restriction enzymes ia a protein that cleaves DNA at specific site. Restriction enzymes also possess a methyltransferase activity. The Nobel Prize in 1978 in physiology and medicine was shared by Werner Arbor, Daniel Nathans and Hamilton Smith for the discovery of restriction enzyme.

## Types of Restriction enzyme

Restriction enzymes are of three types:

1. Type I
2. Type II
3. Type III

### Type I Restriction enzyme

Type I enzymes are complex, multisubunit that cut DNA from far their recognition sequences.

### Type II Restriction enzyme

Type II enzymes are short, palindromic and in the presence of $Mg^{2+}$ they cleave the DNA within their recognize sequences. Type II restrictions enzymes are also involvement of host-parasite interactions.

### Type III Restriction enzyme

Type III Restriction enzymes possesses a sequence specific ATPase activity for DNA cleavage.

## Southern Blotting

Southern blotting was first introduced in 1975 by Edwin Southern. Southern blotting is used in molecular biology to detect DNA. Edwin Southern Kelsar Prize awarded in 2005 for his discovery. Generally, blotting is a technique that transfer DNA from gel to nitrocellulose membarane or nylon membrane. Southern blotting seperates DNA from different cell types by size.

### Procedure of Southern blotting

1. DNA is cut into smaller fragments by using Restriction enzyme.

2. Then, the DNA fragments are separated by gel electrophoresis.

3. After electrophoresis, DNA is transferred to positively charged nylon membrane.

4. The gel is soaked in an alkaline solution to denature the double stranded DNA into single strands.

5. Wash the nylon membranes with perfect Hyb™ plus buffer to block DNA interactions.

6. Prepare fresh probe DNA and labelled with $P^{32}$ alpha-labeled dCTP. Then, detect the probe.

7. The DNA can annealed to the paper on exoposure to heat (80°C). The paper after through washing is exposed to labeled cDNA probe.

Southern blotting is used to identify paternity, thieves, rapists etc.

**Agarose gel electrophoresis**

**Agarose is** isolated from the seaweed genera Gelidium, Gracilaria etc. This technique involves the separation of molecules based on their charge and size, in addition to electrical charge to migrates towards the positive and negative poles. Because, DNA is negatively charged, it is loaded into wells at the negative pole of the gel and migrates toward the positive. DNA is cut with Restriction endonucleses prior to electrophoresis. Smaller fragments of DNA can be moved faster through the gel. Small DNA molecules usally yield only a few bands.

DNA as well as RNA are normally visualized by stainning with ethidium bromide, SYBR Green, GelRed, methylene blue, brilliant cresyl blue, Nile blue sulphate and crystal violet which intercalates into the major grooves of the DNA and fluoresces under UV light.

**Polymerase Chain Reaction**

The PCR is a instrument that copies specific DNA sequence. The polymerase chain reaction can copy segments of DNA. The polymerase chain reaction is used in a various research laboratory and medical institute. PCR is developed by Kary Mulis in 1984. Kary Mulis received Nobel Prize for Polymerase Chain Reaction in 1993.

Requirments: The essential requirements for PCR are listed below-

1. A target DNA (100-35000 bp in length).

2. Two primers (17-30 nucleotides length).

3. Four deoxyribonucleotides (dATP, dCTP, dGTP, dTTP).

4. Taq DNA polymerase.

5. Reaction buffer.

Procedure of PCR

All the PCR components are mixed together and are

1. Denaturation: On raising the temperature to about 95°C for 15-30 seconds, the double stranded DNA gets denatured to produce single stranded DNA.

2. Renaturation: As the reaction mixture slowly cooled to about 54-60°C for 20-40 seconds. This allows the primer to anneal to their complementary sequence in the template DNA.

3. Synthesis: This step usally occurs at 72-80°C. In this step, the DNA polymerase synthesizes a new DNA strain complementary to the DNA template strain by adding free dNTPs from the reaction mixture that are complementary to the template in the 5'-3' direction, for Taq DNA polymerase, the optimum temperature is around 75°C. In practice, 20-30 cycles are usally run, yielding a $10^6$- fold to $10^9$- fold increase in the target sequence.

The thermophilic hot spring bacterium Thermus aquaticus, an organism isolated by Thomas brock by American Microbiologist. T.aquaticus is stable to 95°C. The error rate for Taq polymerase under standard conditions is $8.0 \times 10^{-6}$. Other thermostable DNA polymerases currently used in PCR are Pfu (Pyrococcus furiosus), BstE (Bacillus stearothermophilus) and Tth (Thermus thermophilus).

**Applications of PCR:**

1. PCR is very important tool in the study of evolutionary biology.

2. PCR is employed in the diagnosis of disease.

3. PCR is used in the detection of bacterial infections.

4. PCR is very important tool for identification of criminals.

5. PCR is also used in drug discovery.

6. PCR is used in sex determination of embryos and studies of molecular evolution.

7. PCR has become a major method of analysis and construction of DNA for research and practical purposes.

8. PCR can be used in analysis of gene expression and archaeology, to identify human and animal remains, including insects trapped in amber, and to truck human migration patterns; degraded DNA samples may be able to be reconstructed during the early cycles of PCR.

**References**

1. Roberts, Richard J., and Kenneth Murray. "Restriction endonuclease." Critical Reviews in Biochemistry and Molecular Biology 4.2 (1976): 123-164.

2. Kessler, Christoph, and Vicentiu Manta. "Specificity of restriction endonucleases and DNA modification methyltransferases—a review (Edition 3)." Gene 92.1 (1990): 1-240.

3. Roberts, Richard J., et al. "A nomenclature for restriction enzymes, DNA methyltransferases, homing endonucleases and their genes." Nucleic acids research 31.7 (2003): 1805-1812.

4. Pingoud, Alfred, and Albert Jeltsch. "Structure and function of type II restriction endonucleases." Nucleic acids research 29.18 (2001): 3705-3727.

5. Bagasra O et al. (1992). Detection of human immunodeficiency virus type 1 provirus in mononuclear cells by in situ polymerase chain reaction. New Engl J Med 326, 1385–1391. PMID: 1569974 de Bruijn MH (1988). Diagnostic DNA amplification: No respite for the elusive parasite. Parasitol Today 4, 293–295. PMID: 15463008

6. Eckert KA and Kunkel TA (1993). DNA polymerase fidelity and the polymerase chain reaction. PCR Methods Appl 1, 17–24. PMID: 1842916

7. Meselson, Matthew, and Robert Yuan. "DNA restriction enzyme from E. coli." Nature.; (1968);217(5134):1110-1114.

8. Dussoix, Daisy, and Werner Arber. "Host specificity of DNA produced by Escherichia coli: II. Control over acceptance of DNA from infecting phage λ." Journal of molecular biology 5.1 (1962): 37-49.

9. Hayashi K (1994). PCR-SSCP analysis and its application to DNA diagnosis. Fukuoka Igaku Zasshi 85, 74–77. PMID: 1842918

10. Herschleb, Jill, Gene Ananiev, and David C. Schwartz. "Pulsed-field gel electrophoresis." Nature protocols 2.3 (2007): 677-684.

11. Velculescu, Victor E., et al. "Serial analysis of gene expression." Science 270.5235 (1995): 484-487.

12. Kuspa, Adam. "Restriction enzyme-mediated integration (REMI) mutagenesis." Methods in molecular biology. 2006 ;346:201-209.

13. Higuchi R et al. (1988). A general method of in vitro preparation and specific mutagenesis of DNA fragments: Study of protein and DNA interactions. Nucleic Acids Res 16, 7351–7367. PMID: 3045756

14. Lee CC et al. (1988). Generation of cDNA probes directed by amino acid sequence: Cloning of urate oxidase. Science 239, 1288–1291. PMID: 3344434

15. Mullis K et al. (1992). Specific enzymatic amplification of DNA in vitro: The polymerase chain reaction. 1986. Biotechnology 4, 17–27. PMID: 1422010

16. Ochman H et al. (1988). Genetic applications of an inverse polymerase chain reaction. Genetics 120, 621–623. PMID: 2852134

17. Costello, Joseph F., Christoph Plass, and Webster K. Cavenee. "Restriction landmark genome scanning." Methods in molecular biology. 2002;200:53-70.

18. Bernatzky, Robert. "Restriction fragment length polymorphism." Plant molecular biology manual. Springer Netherlands, 1989. 467-484.

19. Blanc, D. S., et al. "Epidemiological validation of pulsed-field gel electrophoresis patterns for methicillin-resistant Staphylococcus aureus." Journal of clinical microbiology 39.10 (2001): 3442-3445.

20. Wang Y et al. (2004). A novel strategy to engineer DNA polymerases for enhanced processivity and improved performance in vitro. Nucleic Acids Res 32, 1197–1207. PMID: 14973201

21. Arber, Werner, and Stuart Linn. "DNA modification and restriction." Annual review of biochemistry 38.1 (1969): 467-500.

22. Welsh J and McClelland M (1990). Fingerprinting genomes using PCR with arbitrary primers. Nucleic Acids Res 18, 7213–7218. PMID: 2259619

www.ingramcontent.com/pod-product-compliance
Lightning Source LLC
Chambersburg PA
CBHW051934210526
45473CB00006B/2250